How To Make A Scientific Research Poster

By: John Elder

MakeSigns.com

How To Make A Scientific Research Poster

By John Elder

http://www. MakeSigns.com

ISBN: 1495212238

Published By MakeSigns.com
Chicago, IL

Copyright © Graphicsland, Inc.
http://www.MakeSigns.com

All rights reserved. No part of this book may be reproduced or transmitted in any form or by any means without permission in writing by the editor or publisher, except when used by a reviewer in advertisements for this book, or other books or products by the author.

"This book is sold with the understanding that the publisher and author are not engaged in rendering legal, accounting, or other professional services, and is not intended to take the place of such services or advice. If legal advice or other expert assistance is required, the services of a competent professional person should be sought"

--From a declaration of principles jointly adopted by a committee of the American Bar Association and committee of the Publisher's Association

Table Of Contents

Chapter 1....The Two Parts of Making A Scientific Poster

Chapter 2.....Technical Aspects of Making A Scientific Poster

Chapter 3.....How To Make Your Poster STAND OUT!

Chapter 4.....Resources

CHAPTER ONE

The Two Parts of Making A Scientific Poster

INTRODUCTION

Hello! Welcome to "How To Make A Scientific Research Poster", my name is John Elder from **MakeSigns.com** and I'll be walking you through the ins and outs of creating a scientific research poster today.

You've spent so much time, sweat, and tears working on your research for months or even years that the thought of compressing it into a single poster might be daunting for you (and a lot of others!) but it doesn't have to be.

Hopefully after reading this short book you'll have all the tools and information you need to make a great poster.

I mention short, and I mean it. This book isn't going to be very long for two very specific reasons.

One; it doesn't have to be! I can teach you everything you need to know to make a great poster in less than a hundred pages (probably MUCH less than a hundred pages).

Two; you've spent enough time on your research already; you shouldn't have to spend any more time learning how to make a poster than is absolutely necessary.

So with that in mind, I want this book to be a quick and dirty guide that gets you up and running just as fast as possible. Don't worry, I'm not going to scrimp on the info, you'll have everything you need to make a great poster.

And I'll also point you to some free resources online where you can dig deeper and learn even more if you want.

My goal isn't to turn you into an Adobe Photoshop or Illustrator design master, my goal is to give you the tools you need to make a great poster with the least amount of effort, pain and headache.

Here at **MakeSigns.com** we've printed thousands of scientific posters over the years (often at a moment's notice – you'd be surprised how often people leave printing to the last possible minute!) and so we've seen and done it all.

Hopefully I'll be able to impart some of the insights we've gleaned over the years to give you the tools and knowledge needed to make your poster the best that it can be. After all – your research deserves it!

So let's dive right in…

Let's Break This Down

When you get right down to it, there are really just two main areas to focus on when making a scientific research poster.

Think of them as two themes. This book is going to spend an equal amount of time discussing each of those themes, but first things first – we need to take a moment to identify those themes.

So what are they?

The first theme is what I like to call the *technical aspects*. These are things like:

- What software should you use to create your poster?

- What size should your poster be?

- What fonts and font sizes should you use?

- What color schemes should you use?

- Should you use a template or design your own?

- What file formats should you use for charts, graphs, and images and how?

- What is whitespace and what's it good for?

- Bold, Italics, Underlined text – what's appropriate and where?

- What are aspect ratios and why are they important?

- What should you expect to pay for a poster at different places?

- Etc

Nobody really likes to mess with the technical aspects of poster printing, they're sort of like the grammar lessons back in English class; necessary, but not a lot of fun.

But of course, you can't write a great novel without great grammar…and likewise you can't make a great poster without at least an inkling of the technical aspects.

Don't worry though, I'll make it as painless as possible. I'll teach you the technical stuff, but I'll also point you in the right direction to get free templates that take care of most of those technical aspects for you.

And if you work with a decent printing company (like **MakeSigns.com** – not to blow our own horn) then they've got trained staff who should proof every single poster that comes in to make sure the technical aspects are correct.

At MakeSigns.com we make changes if needed, and let you know what's going on every step of the way.

Still, some of you will choose to print your poster on your school or department's poster printing machine or at a small local print shop or the local Kinko's / Fedex - and in each of those cases you probably won't have a knowledgeable design rep looking over your poster to make sure it looks right.

In that case, you're going to need to know this technical stuff before your poster goes off to the printer so we need to spend time discussing those things.

The other theme that we'll be focusing on is *style*.

I like to think of this as Marketing (in fact, I come from a marketing background so this part is right up my alley). Why marketing? Because that's really all your poster is…it's a big billboard advertisement for your research.

It needs to follow some of the same laws of marketing that any advertisement must follow. It needs to stand out, grab attention, motivate, excite…all without looking garish or trying too hard.

Marketing.

The first theme is about how to make a poster. The second theme is about how to make a ***good*** poster. One that does what you want it to do…which is to get your research noticed.

Seem pretty straight forward? It is! So let's get started with the technical aspects.

CHAPTER TWO

Technical Aspects of Making A Scientific Poster

Let's talk about the technical aspects of making a scientific poster.

WHAT SOFTWARE SHOULD YOU USE?

First off...what software should you use to make a research poster? There are really a couple of different options (not really – there's only one *real* option but for the sake of choice and being complete I'll talk about the other so-called "options").

You'll need some sort of software to help you build your poster. The options are basically:

- ***Microsoft PowerPoint***
- Adobe Photoshop
- Adobe Illustrator
- Adobe InDesign
- Corel Draw
- Gimp
- LaTex
- Open Office

Microsoft PowerPoint is the program you're going to want to use. Why? Because in my opinion, that's what everyone else is going to use, that's what everyone else has always used, and that's what's going to be the absolute easiest program for you to use.

The drawback is that it can limit your options from a purely design related standpoint. If you have very definite ideas about how your poster should look, and you have massive design skills, then you're going to want to use some professional design software like Adobe Photoshop, Illustrator, or Indesign etc.

Of course, those programs cost thousands of dollars and have a MASSIVE learning curve for people who have never used them. Again – stick with PowerPoint.

Chances are, you probably already have PowerPoint on your computer, or on a computer in your department. If not, it's a quick and easy download.

Yes PowerPoint is boring and dull – but that's the point. You *want* boring and dull. You want easy and straight forward. This is not an area where you need to blaze trails. Stick with the tried and true here.

It's not just about ease of use – there are tons of free PowerPoint templates that you can instantly download and use to handle all the messy technical aspects of making your poster.

It really becomes a matter of just browsing through a stack of templates online to find one that suits your fancy, downloading it, opening it, and then pointing and clicking around on the different areas of the template and pasting your research text in.

Why make it any harder than that?

So that's the track we're going to take throughout the rest of this book and it begs the question…where can you find free PowerPoint templates?

There are basically two options. First, you can head over to Google and run a search for "Free Scientific Poster PowerPoint Templates". You'll find a bunch of listings (including ours).

The second option is to skip the searching and head straight to our website, MakeSigns.com. We have a bunch of free templates ready to browse and download.

Here's a direct link:

http://www.makesigns.com/SciPosters_Templates.aspx

Our templates all come in a variety of sizes, colors, and aspect ratios so you can download the template to fit the poster size that you had in mind. Each template gives you the option of selecting different sizes and aspect ratios so you can find the exact size to fit your needs.

Apart from our list of templates, we also offer a free option for universities and organizations where we'll list your templates on our website.

So if you are a school who wants their students to use a specific set of templates, we'll host those templates for you (often those are custom templates with your logos and color schemes).

Just contact us if you are an organization who would like to make your templates available to your students or colleagues on our site (or would like us to build some templates for you).

Oh – I almost forgot. If you just don't have access to PowerPoint, and don't want to spend the money to purchase it, you can get by with OpenOffice; which is a free open source alternative to the Microsoft Word suite of tools.

OpenOffice has its own knockoff version of PowerPoint called "Impress" which should work ok for you.

I believe it will work with PowerPoint file types so you can use all the free PowerPoint templates with it.

Like I mentioned, true believers and hardcore designers will roll their eyes at the mere thought of using PowerPoint...and maybe with good reason. But let's face it; you're not a hardcore designer and neither am I.

When I make a scientific poster I want quick, easy, and decent looking. PowerPoint fits the bill perfectly. So why make it harder than it has to be?

WHAT SIZE SHOULD YOUR POSTER BE?

Before we get into trickier sizing areas like aspect ratios and confusing things like that, we should talk a little bit about basic poster sizes.

So what size should your poster be?

The answer is pretty simple; it doesn't really matter! Sure there are some standard sizes that you're probably going to want to stay within, but the final size is up to you.

That being said, there are some things to keep in mind and that's what I'm going to talk about in this section.

First of all, there are some limits that PowerPoint will impose on you...well *sort of*. What do I mean?

Well, PowerPoint has a max page width of 56 inches. But we can get around that if you want. You'll just have to use a smaller sized template and then a decent printer can resize it based on the aspect ratios you have chosen.

Here at MakeSigns.com the largest poster we can print is 48 x 120, just fyi.

So basically your poster can be either square or rectangular and have either a portrait or landscape orientation.

Portrait vs. Landscape

One choice you DO need to make is whether your poster will be portrait or landscape oriented. What's the difference? I like to think of portrait as tall and landscape as wide. Here's a couple of pictures showing the difference:

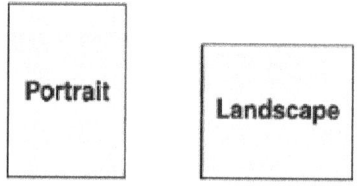

When you design your poster in PowerPoint, your page layout will determine if your poster is portrait or landscape.

Don't be concerned with which number is the "height" and which number is the "width". If you order a 36x48 poster and your PowerPoint page setup is portrait, then your poster will be printed portrait.

If your PowerPoint page setup is landscape, then your poster will be printed landscape. Most posters are landscape.

So what are some common sizes? Here's a list of some common sizes:

36x48, 36x54, 36x60, 36x72, 42x56, 42x63, 42x84, 48x72, 48x96.

Most poster conferences will supply information about recommended poster size. The most popular size for a poster presentation board is 48" **high** x 96" wide (4 by 8 feet).

Many researchers feel they need a poster that same size. But 48x96 is the size of the **board, not the poster**. Your poster doesn't need to cover the whole board. A 36x72 or 42x84 poster fits nicely on a 48x96 board.

MakeSigns.com provides templates in several different aspect ratios that can be printed in all the popular poster sizes. You'll see a list of those if you click on any of the templates on our template page:

http://www.makesigns.com/SciPosters_Templates.aspx

In fact, you can choose to download any of our templates in any of those sizes just by clicking that size's link on the template page.

So what happens if you've already designed your poster in PowerPoint or some other piece of software and you need to change the sizes to fit the specification of your printer? You can do that!

In fact, we've built a free little online calculator that will determine the size you need to change it to based on your original dimensions to fit whatever aspect ratio is needed. You can find the tool here:

http://www.makesigns.com/SciPosters_PageSizeConverter.aspx

We've also built a free little tool that lets you plug in your original dimensions and it will tell you all the different sizes that will work with your current aspect ratios. You can find it here:

http://www.makesigns.com/SciPosters_PageSizeCalculator.aspx

Both of those tools are completely free, there's nothing to sign up for, you don't have to enter your email address or anything; heck – you don't even need to use us as your printer. They're just on our website for anyone to use.

We built them because in all of poster printer-dom, page sizing and aspect ratio stuff seems to confuse the most people. I get that, it confuses me too!

But with these little tools you can sort things out fairly quickly. But if you're still confused, give us a shout and we'll walk you through it.

You can reach us on our website, via email, via our online chat function during business hours, or by telephone toll free at 1-800-347-2744.

And again, we're happy to help answer questions even if you don't ultimately use our printing services.

If you DO need to resize your poster, those tools will tell you what size your poster can be changed to based on the original aspect ratios, etc.

But to actually RESIZE your poster in PowerPoint, follow these steps:

1. Open your existing PowerPoint slide.

2. Select everything on your slide by pressing Ctrl + A.

3. Cut everything from your slide by pressing Ctrl + X.

4. Go to the Design tab and select Page Setup.

5. A dialog box will appear similar to the one on the right. Enter the width and height of

the page. The slide orientation (portrait or landscape) will automatically adjust based on the page sizes entered.

6. Click OK.

7. If the rulers are turned on (on the View tab check the Ruler box), you will be able to see that the slide is now the size you entered.

8. Paste all of the items back on the slide by pressing Ctrl + V.

9. At this point, you will most likely have a lot of extra space on your slide, or have a lot of your content hanging off of the slide boundaries. Be sure to rearrange your content so that everything is displayed nicely on your newly resized slide.

The page size of your PowerPoint document must be the same aspect ratio as your final output size. Your file's page size need not be the same as the final output size, but must be proportionate.

If, for example, you're going to order a 36x72 inch poster, your PowerPoint page size can be set to 24x48 or any other proportionate page size.

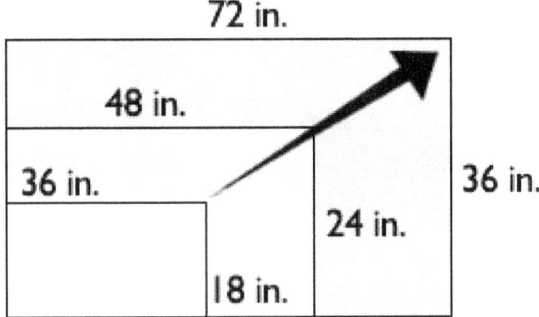

Many customers get frustrated when we tell them that we can't print their poster to a size that is not a proportionate size.

We aren't trying to be difficult, we're just looking out for you! Printing a poster to a size that is not proportionate to the file page size results in poster contents being distorted, squished, stretched, and just plain old crummy looking. Just take a look at some of the examples below.

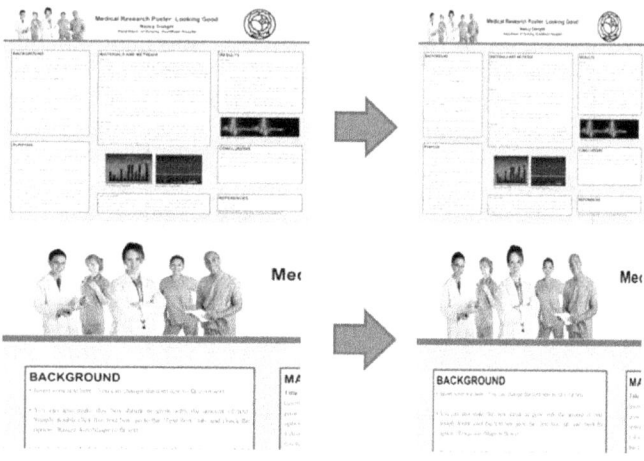

Finally, I want to also mention that we have a tutorial posted on our website that goes into greater detail about sizing and different size issues. You can check it out for free here:

http://www.makesigns.com/tutorials/poster-sizing.aspx

WHAT FONTS AND FONT SIZES SHOULD YOU USE?

Generally speaking, the template that you use will probably already have the font sizes and typefaces chosen for you.

You'll only need to copy and paste your information into the specific areas of the poster where you want them to show up.

But if your template doesn't have that set up already, or if you're designing your own template or something crazy like that, here are some general guidelines for text sizes.

These aren't set in stone – by any means – but are generally accepted sizes for the main areas of your poster. Of course, you can feel free to tinker with these sizes to fit your exact needs.

Generally there are going to be three or four areas in your poster, and each of those areas will probably need a different font size just to differentiate each area and set them apart so the eye can make out the different sections of your poster.

If the font size was the same throughout your entire poster, the eye wouldn't know where to start reading and it would look like a jumbled mess.

The main sections of your poster will likely be:

- The Headline - this is where the Title of your poster goes.

- The Headline Sub-Head - this is directly under the Title and usually where you put the author's name – ie. your name.

- Section Sub Headlines – These are the bold headlines of each paragraph section.

- Body Text - This is the main paragraph styled content of your poster

- Captions – These are found directly underneath any charts, graphs, or images you might have.

That's pretty much it! So the question becomes; what size text should each of those sections have.

Here's a quick List:

- Headline: 85 point, possibly bold

- Headline Subhead: 56 Point, possibly italic

- Section Sub Headlines: 36 Point, probably bold

- Body Text: 24 Point

- Captions: 18 Point, possibly italic

You can see the proportions I use throughout the list. The headline is big and bold, it's meant to stand out and draw attention.

The Section Sub-Heads are meant to do the same thing, stand out and draw attention to each section but at the same time need to be smaller than the main Title of your entire poster.

Finally, the body text should be smaller than the Section Sub-Head text and Captions should be the smallest text of all.

This is all pretty much common sense. Does your Title Headline have to be 85 point? Of course not. But if you make it 75 point, change all the other sizes downward in the same general proportion and you should be fine.

Like I said earlier though, this is pretty much a moot point because most of your templates are already going to have the font text sizes already baked in.

And what fonts should you use? It doesn't really matter all that much which fonts you use, so long as you follow a few simple rules.

1. Don't use wacky fonts - stick with traditional fonts like Arial, Helvetica, Times New Roman, etc.

2. Don't use drop shadow for your text.

3. Use Sans Serif fonts, not Serif because they look better at a distance:

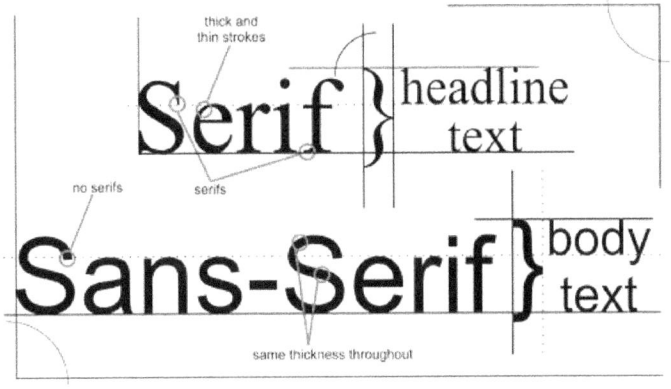

4. Stick with one or two *types* of fonts throughout your entire poster. Here's a list of good pairs:

MakeSigns.com

Helvetica / Garamond
Caslon / Univers
Futura / Bodoni
Garamond / Futura
Gills Sans / Caslon
Minion / Gill Sans
Myriad / Minion
Caslon / **Franklin Gothic**
Trade Gothic / Clarendon
Franklin Gothic / Baskerville

TEXT ALIGNMENT

While we're on the subject of fonts and text, I want to throw in a quick tip. Your text should all be "Left-Align" text (sometimes called "Left-Justified").

Headlines can be centered, but the actual body text throughout your poster should almost always be left-aligned.

Some people like to center all their text within each little sub-section. That's probably a bad idea. It tends to just look sloppy.

Other people like to "Justify" the text of each section. By Justify, I mean make all the lines of the paragraphs the same width, just like a newspaper. So instead of this:

After further testing, I discovered that the test subject showed signs of decay.

...you get something like this with justified text:

After further testing, I discovered that the test   subject   showed   signs   of   decay.

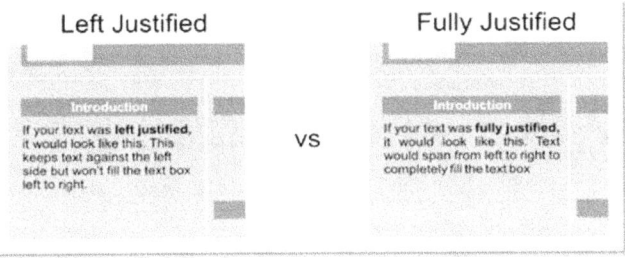

Your gut is going to tell you that justified text will look good because your poster is set up with lots of little sub-sections, or squares and rectangles...and fitting the text to those squares and rectangles will give the entire poster a clean look.

The reality is that justified text ends up being harder to read on posters. The stretching of those lines to fit the square can end up with some real wacky looking text that's just hard to read.

Justification can change within different versions of PowerPoint. Using full justification on narrow text boxes will leave you with open spaces or "rivers" in the text (see below).

(PowerPoint Windows)		(PowerPoint Mac)	
Fully Justified	Left Justified	Fully Justified	Left Justified
Your gut is going to tell you that justified text will look good because your poster is set up with lots of little sub-sections, or squares and rectangles...and fitting the text to those squares and rectangles will give the entire poster a clean look.	Your gut is going to tell you that justified text will look good because your poster is set up with lots of little sub-sections, or squares and rectangles...and fitting the text to those squares and rectangles will give the entire poster a clean look.	Your gut is going to tell you that justified text will l o o k g o o d because your poster is set up with lots of little sub-sections, or s q u a r e s a n d rectangles...and fitting the text to those squares and rectangles will give the entire poster a clean look.	Your gut is going to tell you that justified text will look good because your poster is set up with lots of little sub-sections, or squares and rectangles...and fitting the text to those squares and rectangles will give the entire poster a clean look.

So follow the rule of thumb, keep your text "<u>Left-Justified</u>".

WHAT COLOR SCHEMES SHOULD YOU USE?

Let me start right off the bat by saying that I am NOT a designer. I mean, I can cobble together a logo or a website design with the best of them, but colors confuse me.

There's this thing called a color wheel, and it does a thing with the thing, and a this and a that…and I don't understand it at all!

But I can point you in the right direction so that you can figure it out, and I can give you some general rules of thumb to make sure your poster doesn't look hideous.

Like font sizes, there's a pretty good chance that the template you download is going to already have the color scheme selected and there won't be anything you need to do so long as you're happy about how the template looks.

But if you want to mess with the color scheme, PowerPoint makes it pretty easy to do, and that IS something I can tell you about.

But first things first, let's try to talk about color combinations. Since I'm so bad at colors, I asked our design team to point me in the right directions and they provided a bunch of great resources.

First, they showed me this Color Combinations PDF that you can download and check out from our website here:

http://forums.makesigns.com/index.php?app=core&module=attach&section=attach&attach_id=2

It's specifically designed to work with PowerPoint.

One of the nice things about using the templates that we give away on our site is that they all have multiple color schemes baked into them (and the color schemes were picked by our designers – not me!).

To toggle between different color schemes in PowerPoint, just open the template file that you downloaded from us, and click on the "Design Tab" there at the top of the screen. There on the right side of that menu you should see a "Color" tab, click it.

A drop down box will pop up with all the different color combinations that are available in that template. Just hover your mouse over each one and your template will automatically update to that color scheme…that's a good way to preview them all and find the one you like.

When you find the one you like the best, just click it to select it.

That's all there is to it!

Of course, that's only good if you're using one of our free templates. If you're using someone else's template, or if you're designing your own, then you need to pick the colors manually.

The best way to do that is to use an actual color wheel. There are tons of color wheel websites with free color wheels. Here are just a few to choose from:

- https://kuler.adobe.com/create/color-wheel/

- http://www.colourlovers.com/

- http://colorschemedesigner.com/

Personally I like the last one the best, colorschemedesigner.com because it's pretty simple. You just click somewhere on the big wheel around whatever main color you want your poster to look like.

So if I want blue, I'll click on the wheel close to the type of blue I want. The site will then show you a bunch of blue-type colors that go with that particular shade of blue. Etc.

There's also a sort of menu at the top of the screen that lets you choose different color options such as mono, complement, triad, tetrad, analogic, and accented analogic.

Now I don't know what any of those things means from a color science point of view, but you can mess around with it if you're interested.

What I CAN tell you about color is that the best thing to do is follow the KISS principle. KEEP IT SIMPLE, STUPID!

This is not your chance to let out your inner designer. This is not your chance to show the flourish of style that you've always kept hidden away.

The rule of thumb is to keep colors basic. Don't use garish colors, don't use loud colors. The purpose of your poster is to get your research noticed…but not by making your poster look like a clown.

And don't use tons of different colors. Two or three colors is about all that you want to put in there. Any more than that and things start to look confused.

Text Color

And while we're at it, let's talk quickly about text color.

Text should almost always be black, or a very dark color, set against a very light background. Like the words on the page of a book.

People sometimes try to use a dark background with WHITE text because they think that will make their text stand out. DON'T DO THAT!

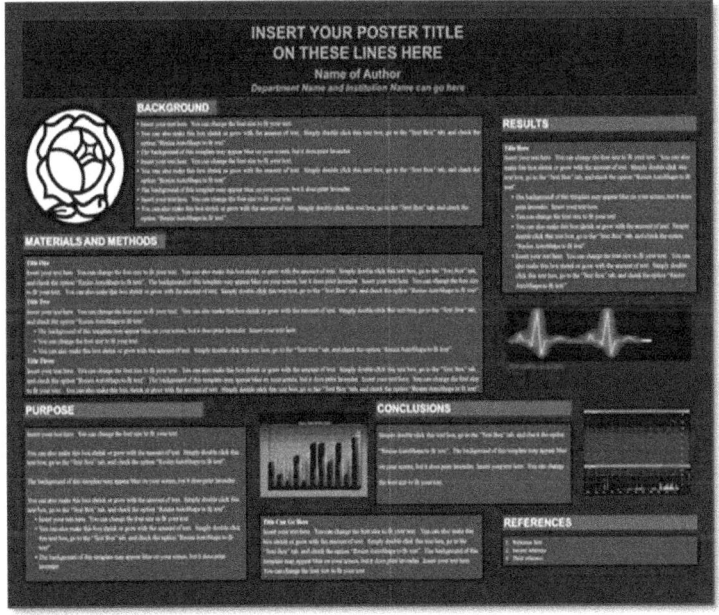

Isn't that hard to read?! I'm a marketing guy...and one of the very first things you learn about print advertising is to not ever use white text on a black or dark background.

It's has been tested out the wazoo...there's something about our eyes and how we read that makes deciphering white text on a dark background tiring to our eyes. We just don't like it, subconsciously.

Let's play with the colors of the poster above and then compare the differences and you can tell me which one is easier to read...

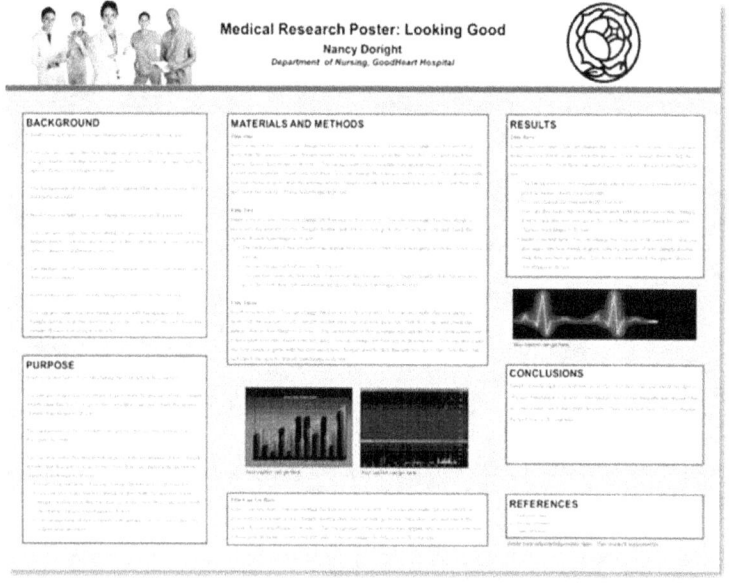

No contest, right? So stick with dark text on light backgrounds...almost always.

Background Color

Finally, what color should you use for the overall background of your poster. Again, it doesn't really matter as long as it's light and doesn't clash with the text.

Sometimes people try to use busy images for their background. Someone presenting a poster about the solar system might try putting a full color picture of a planet or the solar system as their background.

That's generally a bad idea. Don't try to be cute, keep it simple and easy to read.

Here's an example of what I'm talking about. Just how easy do you find reading this poster?

And here's that same poster without the crazy background. Just removing the background makes it so much easier to read! (sure it's still not *great*, but it's certainly an improvement on the crazy background version above).

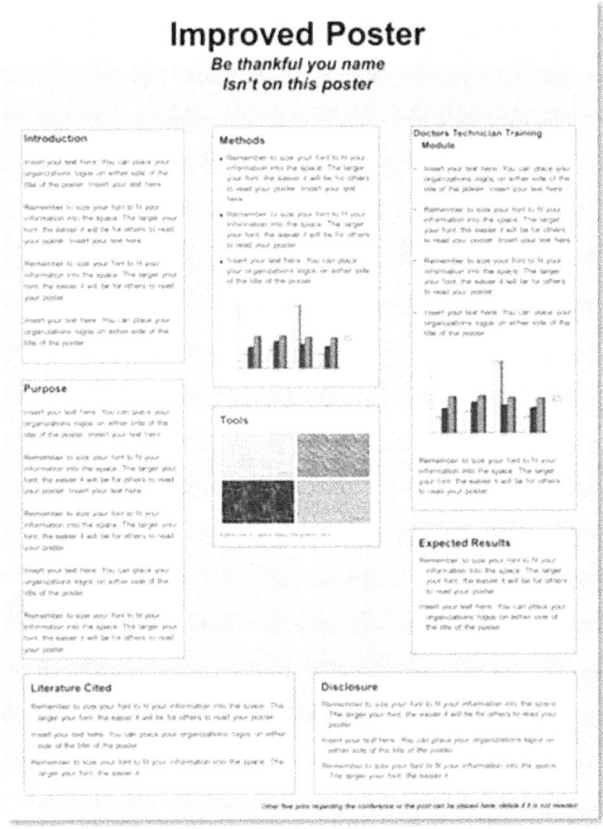

Background images, unless they are very subtle, tend to make a poster harder to read. Sure there are exceptions, but generally speaking you're going to want to stay away from them and go with simple light toned colors.

The template you download and use will usually already have this set for you.

In conclusion, colors can be tricky.

It's crazy how something so simple can really destroy your entire presentation... but it can.

Stick with simple color schemes, don't use loud or busy backgrounds, and stick with traditional black or dark colored text on lighter backgrounds and you should be ok.

Or better yet, stick with one of the ready-set color schemes that comes with your template!

SHOULD YOU USE A TEMPLATE OR DESIGN YOUR OWN?

I guess this section should have really gone closer towards the beginning of the book because if you've been paying attention at all so far, then you probably already know how I feel about this and we really don't need to spend much time on it.

In my opinion, you should always use a template. In fact, unless you have a design background and lots of free time, there's really no reason NOT to use a template.

I've mentioned our own repository of free templates online that you can grab right here:

http://www.makesigns.com/SciPosters_Templates.aspx

I think they're pretty good, but if you don't find exactly what you're looking for then a quick Google search for Free Research Poster Templates should do the trick.

So the short answer to the question above is; yes you should absolutely use a template!

But while we're on the subject of templates and general look and feel of your poster, I thought it might be a good time for a tip.

Sometimes it's hard to know where to start, even when choosing a pre-made template. Heck, you still have to PICK the template!

Some people, myself included, have a hard time visualizing what the poster should look like. It's hard to start with a blank slate, especially when it's this important.

One thing I like to do is browse other people's posters for inspiration. You shouldn't exactly copy someone else's look or design, but browsing is a great way to get ideas, to point you in the right direction, and to get the train moving down the track.

The Internet is a great tool for this sort of thing.

One great place to start looking is Flikr, the photo-sharing site. They've got a group called "PimpMyPoster" where people upload their poster designs and the good people of the Internet comment on them, giving suggestions, telling you what looks terrible and how to fix it.

You can check it out here:
http://flikr.com/groups/pimpmyposter

Regardless of whether or not you upload your own design, you can at least browse through the listings of other people's posters for inspiration.

Maybe the layout of one poster strikes your fancy and the color scheme of another one looks good. Whatever. It's a great place to get ideas and start the juices flowing, especially if you don't know where else to start.

Another place to look is f1000.com They've got several thousand posters listed on their website and you can browse to your heart's content. Check out the list here:

http://f1000.com/posters/browse?docTypeSearch=Poster

Like I said, absolutely don't copy someone's design…but there's nothing wrong with getting a little inspiration from someone else's poster.

MakeSigns.com

WHAT FILE FORMATS SHOULD YOU USE FOR CHARTS, GRAPHS, AND IMAGES?

The next thing to talk about is file formats, both for charts, graphs, and images…but also for your poster in general.

When saving your overall poster file, it should really be saved as a PowerPoint Presentation file – .PPTX file (that is, if you're using PowerPoint).

A good printer can usually take a .PDF file and work with it as well, but they'd probably rather have a PPTX file (or a .PPT file if you're using the earlier versions of PowerPoint).

But more importantly, what file formats should you use when adding charts and graphs and images to your poster. After all, those can be an incredibly important part of your poster and you want to make sure that they show up nicely on the finished poster.

There are TWO main considerations when it comes to images; file format and resolution. Let's talk about resolution first.

Whenever you add any sort of image to your poster, you want the file to be as high a resolution as possible. I'm talking DPI, or "Dots Per Inch".

A lot of people tend to get into trouble here because they use images off the Internet, and those images are usually low DPI images of around 72 DPI.

A good image resolution for a scientific poster is anywhere from 150 to 300 DPI (but generally not more than 300).

The best thing to do is start out with a high quality image. Typically a photo you capture from a web page will be a low resolution image that may look pixelated when you "zoom in" or when the poster is printed

The other thing to consider when it comes to images is file format. When saving images, be they pictures, charts, or graphs; you need to decide which type of file to use. Common image file formats include: .jpg, .gif, .png, .bmp

So which is best? It depends on the image.

Generally speaking, you'll want to stick with either .jpg or .png but the choice between them will deal with the type of image you're using.

For charts and graphs (line art, etc.) you should copy and "paste-special" into PowerPoint as an Enhanced Metafile or an Excel Object.
This will keep the vector format of the chart. Only use .png if this doesn't work for whatever reason. Pictures can be either .jpg or .png.

If you use a professional poster printer like MakeSigns.com, our people will proof your poster before it's printed and if we find any images that won't look good, we'll stop everything and call you to see if you have any better images that we can swap in.

In fact, that happens very often and is one of the main reasons to have your poster professionally printed and not just take it down to Kinkos and hope it turns out ok.

Though if you absolutely have to use Kinkos or your department's old printer machine, here's a tip to help you determine how good your images (and the poster itself – for that matter) will look when printed…

PowerPoint has a "zoom" feature that lets you zoom in and get a pretty good idea how your poster will look when printed.

To use it, you need to know the size of your template (the PowerPoint page setup size of your file), and the size that you'll eventually want to print your poster as.

Let's pretend that your template size is 24' x 32' and that you plan on printing it as a 42' x 56'.

The trick is to take the vertical dimension of the print size and divide it by the vertical dimension of the page setup size in your power point. So in this example, take 42 ÷ 24 = 1.75

Now take that answer and multiply it by 100 to get your viewing percentage. So 1.75 * 100 = 175%

I mentioned PowerPoint has a zoom feature. To use it, click the "View" tab at the top of the screen in PowerPoint, then click the "Zoom" tab.

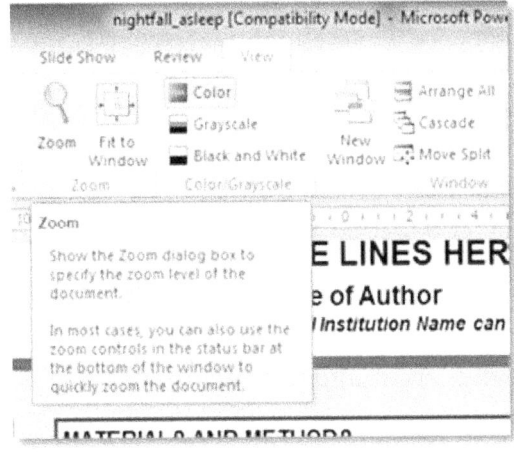

A little Zoom box will appear with a bunch of pre-set options. Ignore them, and instead type in 175 in the little percentage field and click Ok.

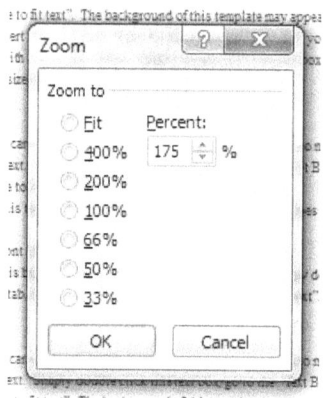

This will ZOOOOM your poster WAAAY in, and may look a little weird. But you can navigate around to the different images and see what they'll look like printed.

Stand a few feet back from your computer monitor and give it a look. Are the images crisp? Are there weird fuzzy lines? If so, you've got a problem and probably need an image with a higher DPI.

IMPORTING IMAGES INTO POWERPOINT

To get your images on to your actual poster you have a couple of options; importing or cutting and pasting.

Importing Images

1. Place the insertion point at the position in your document where you want to insert the picture.

2. On the Insert menu, point to Picture. This will open the Insert Picture dialog box, which is similar to the Open Office Document (or Open) dialog box for opening documents.

3. In the Insert Picture dialog box, locate and select the graphics file you want to import. You can import graphics files in a wide variety of formats—for example, files with the extensions .bmp, .wmf, .gif, and .jpg. To see preview images of your graphics files, click the More Options button in the upper right corner and then select the Small, Medium, Large, or Extra Large view from fly out menu.

4. Click the *Insert* button.

Cutting and Pasting Images

If you have a graphics on another document that you'd like to use on your PowerPoint poster, you can use cut and paste. To do a basic cut and paste simply follow these instructions:

1. Select the graphics in the other program, and from that program's Home tab or Edit menu, click Copy (or Ctrl+C).

2. Place your cursor at the approximate position of your poster where you want to insert the picture.

3. On PowerPoint Home tab, click Paste. (or Ctrl+V)

If you'll need to edit the item in its native application (by double clicking, if available) after it is pasted in PowerPoint, use the above instructions to paste the item in place.

Otherwise, we recommend using the Paste Special feature so you have more control over how your graphics look.

Pasting With The "Paste Special" Function

Go to the Home tab, click the down arrow below "Paste" and select "Paste Special" (you can also use Ctrl+Alt+V).

Once you've done that, you can use the explanations below to choose the best option from the pop-up box. This is a general guide that may vary slightly between the different versions of PowerPoint.

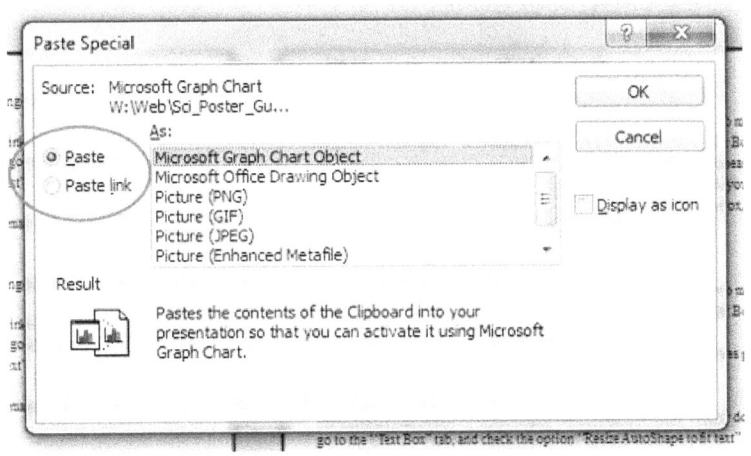

- Items pasted as a Microsoft Office Graphic Object will insert as a graphic object that can be edited in PowerPoint.

- Items pasted as a Picture cannot be edited. Depending on the type of object, the result will be the same as a pasted enhanced metafile or it will turn into a jpg (bitmapped picture).

- Items pasted as an Enhanced Metafile cannot be edited in their native application. They can only be ungrouped to dumb objects in PowerPoint. Ungrouping will cause charts, graphs or vector objects to split up into 100s of pieces.

CHARTS AND GRAPH TIPS

Chances are, your poster is going to have at least one or two charts or graphs. Here's a couple of tips for using them on your poster.

Tip1. Avoid using pattern fills in charts and graphs. Patterned striped fill may look fine on your screen (Figure A), but when it is blown up to the full size, the pattern will shrink and won't be visible (Figure B).

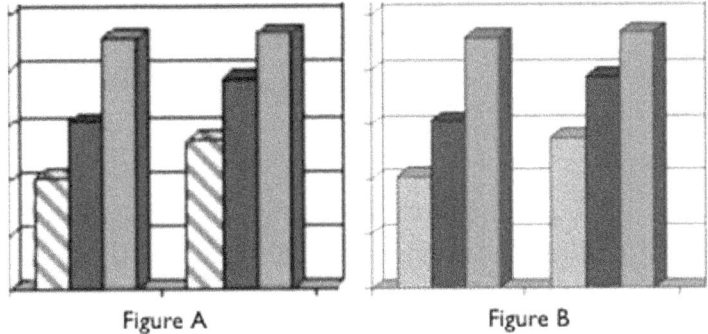

Figure A   Figure B

If you absolutely must use a pattern, here is a workaround you can apply.

- Once the chart is placed on your poster in the right position and at the correct size, click the left mouse button on the chart so that it is selected.

- Next press Ctrl + X (to cut). This process will put your chart onto the PowerPoint clipboard.

- Then press Ctrl + Alt+ V for Paste Special. When the dialog box comes up pick the option Picture (PNG) and press ok. This will convert your Excel chart into a graphic. (WARNING: You will not be able to edit your chart once it's converted to an object.)

- Once this is done, position your chart in the poster. After you have done these steps the stripes in your chart will blow up proportionality as seen in Figure A.

Tip 2. Remove gray backgrounds and gridlines from charts. The gray can make things hard to read, and no one really cares how exact the data is on the graph – they just want a general idea...so gridlines are overkill and can look too busy.

Tip 3. Remove any legends, and reduce the number of tick marks to as few as possible. Legends take up precious space (and you can always explain the legend if someone asks), and too many tick marks on the axis can be hard to read and look cluttered. Remember, people don't really care about exactness at this point, they just need to understand the general data that you are trying to convey.

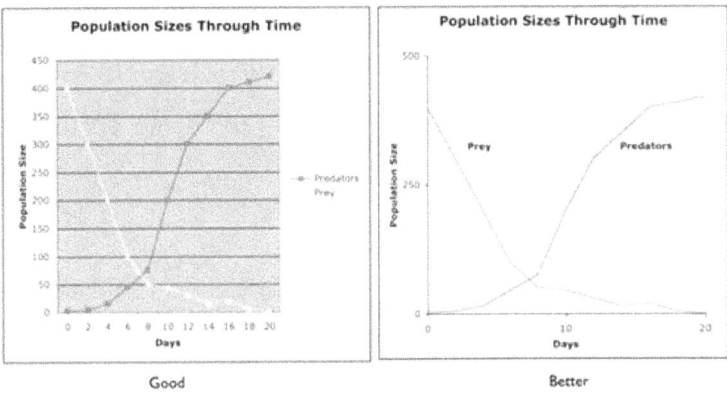

Good　　　　　　　　　　　Better

WHAT IS WHITESPACE AND WHAT'S IT GOOD FOR?

Moving right along, I want to talk very briefly about whitespace. First off, what is it? Basically it's just empty space around the different components of your poster.

By components, I mean a block of text, an image, a graph, a chart, or any other thing that you've sectioned off in some way.

One problem we see often is when people try to cram so much stuff into their poster that there's not enough whitespace.

This results in a poster that's very hard to read because the eye just doesn't know where to start or how to navigate the poster. It also makes the poster feel cluttered.

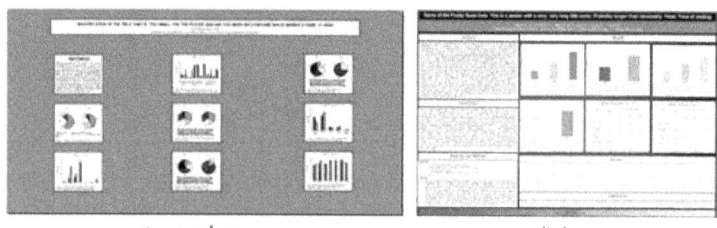

too much space   too little space

I'll talk about this more in the next chapter when I discuss how to make a *good* scientific poster, but I did want to at least introduce the concept of whitespace before we talk about how to use it.

For now, just try to remember that whitespace is our friend. Whitespace can sometimes be more effective in drawing attention to a particular section than the headline of the section or the data itself.

So keep your stuff spaced out. Keep enough space between different sections so that it's easy to tell that there are supposed to be separate sections!

More on whitespace later…

BOLD, ITALICS, UNDERLINED TEXT – WHAT'S APPROPRIATE AND WHERE?

Generally speaking, there are very specific parts of your poster that need very specific type of text, and you probably already have a pretty good idea of where those places are.

For instance, you know that your title (main headline) and section head's should probably be BOLD text.

Key data or phrases might be underlined or maybe both bold and underlined.

Captions for charts and graphs as well as images can sometimes be italic.

For the most part though, the text of your poster should be regular type, not bold or underlined or italics.

Your sub-heads don't need to be both bold AND underlined. Often people feel they need to draw attention to a sub-head by using underlined text and it's often a bad idea.

Trust me, the use of whitespace does better to draw attention to a section than underlining the headline. Of course, you should bold the text…but bold and underline, or bold and italics is probably overkill.

I say probably, because there are no hard and fast set in stone rules. But generally speaking, we want to keep our KISS rule in mind (Keep It Simple Stupid!).

Judicious use of bold, italics, and underlined text outside of the traditional areas (headlines etc) may be appropriate for drawing special emphasis to something, but don't go overboard.

Clearly defined sections, set apart by judicious use of whitespace will do more to draw attention to your data than anything else.

Simple and elegant is usually best, and bold text is like screaming…which is rarely elegant.

WHAT ARE ASPECT RATIOS AND WHY ARE THEY IMPORTANT?

Ok, I suppose it's time to talk about aspect ratios. Believe me, I don't really want to talk about it anymore than you want to learn about it!

It's just one of those things. So let's try to get through this as painlessly as possible.

We've already sort of talked about aspect ratios earlier when we talked about template sizes.

Basically an aspect ratio deals with the proportions between a thing's width and height. Or, more specifically, it's the *ratio* of an item's width to its height. Usually it looks like two numbers separated by a colon, like 16:9 or 4:3.

This is important if you want to resize your poster, you'll need to know the correct size based on the original aspect ratios (or more likely, you'll just use our free online calculator that I mentioned earlier).

So a template that is 24x36 has a 2:3 aspect ratio.

If you want to figure out what a new size should be based on that aspect ratio, you would take the original height divided by the original width, and then multiply that by whatever new width you'd like to have…and that will give you the new height the poster should be.

So if you wanted a width of 64…the new height should be:

$$36/24 = 1.5$$
$$1.5 \times 64 = 96$$

So your new dimensions would be 64 x 96 and that would be in proportion to your original template.

Yeah… you aren't mistaken, this stuff really is mind-numbing. You really don't need to know it, I just added it to the book for completeness sake…just in case you want to know about it.

If you need to resize your poster or determine which sizes your poster can be converted to based on its aspect ratio, save yourself a headache and just use our free online tool here:

http://www.makesigns.com/SciPosters_PageSizeCalculator.aspx

WHAT SHOULD YOU EXPECT TO PAY FOR A POSTER AT DIFFERENT PLACES?

Alright! We've made it through most of the technical aspects of poster printing. That wasn't too bad, was it?

Before we get into the next chapter and talk about making a *good* poster, I want to spend just a minute or two talking about poster printing prices.

Yes, I'm a little biased because I work for MakeSigns.com and we print posters. But I can still give you the facts and let you decide where to get your poster printed.

Before we talk prices, I do want to reiterate something I've touched on throughout this book. I won't harp about it anymore, but it's important so I want to make a point of it even if it seems self-serving.

You really have two options when it comes to printing; you can pay a poster printing service like MakeSigns.com or you can take it down to your local Kinkos or other local general printer.

I highly recommend MakeSigns.com or some other specific scientific poster printing company. Why? Because the kids working down at Kinkos don't know a scientific poster from a convenience store poster and they don't care.

They don't care that it represents all your hard work, that it's important, and that it needs to look perfect. *We do.*

They haven't printed thousands of scientific posters day after day after day for years. They don't know the common problems that people have and how to quickly fix them.

And they don't care that your conference is in two days and YOU NEED THIS DONE NOW (and done well!).

We do; and any other scientific poster printer place should too. We're going to go over your poster to make sure it's perfect, and we won't print it till it is. Those other places just aren't going to do that.

And now I'll talk about prices (and really knock your socks off when you see how much cheaper we are than those other places).

So what can you expect to pay to have your poster printed? Here's a comparison:

http://www.makesigns.com/SciPosters_PriceComparison.aspx

Prices for a 36 by 60 inch poster

MakeSigns.com	**$53.99**
Poster4Research	$62.50
Cp-digital (PosterPresentations)	$85
PosterSession (MegaPrint)	$109
ePingo.com	$142.50
FedEx/Kinkos	$180
Scifor.com	$204.95

(Online prices as of Jan 10, 2013)

I think the choice is obvious!

CHAPTER THREE

How To Make Your Poster STAND OUT!

Ok! So we're passed the boring technical aspects of making a scientific poster...to sum up...use a pre-made PowerPoint template! Now we can dive into the fun stuff!

From here on out I'm going to talk about how to make a *good* scientific poster. We'll look at it from a design and layout point of view, but also from a marketing point of view.

After all, the entire point of your poster is to get your research noticed... so you'll want to pay particular attention to this chapter!

WHAT IS A SCIENTIFIC POSTER – AND WHAT IS ITS PURPOSE?

So let's break it down. There are two ways to look at your poster, and both are important. The first is to look at it like a billboard that needs to grab attention; the second is to look at it as an illustrated abstract of your research.

We'll talk the billboard thing in a moment, first I want to talk about it from the abstract point of view.

So yeah, essentially your poster is an illustrated abstract of your research. That implies two things…illustrated, and abstract. Woah!

ILLUSTRATED

Your poster should definitely have either images, charts, or graphs…or some sort of combination of all three.

I don't think that will come as any shock to you. People tend to go overboard with text and skimp on images and that can be a mistake.

Posters are a drive-by type of thing...conferences are full of people who are mingling...mobile... wandering around a huge room with tons of posters everywhere. Many of them are just there for the free drinks. Keep all of that in mind.

What do people in motion focus on? Images.

You're going to want to pull people in with your title (headline) and your images. Otherwise why would they bother to stop at all?

Sure the title is technically text...but it's also sort of like a big attention grabbing image as well...sort of (go with me on this).

The point is; the body text of your poster is important...the people who are really interested in your work are going to stop and read it... but everyone else is going to give you a few split seconds of eyeball time...and those eyeballs are going to drift from your poster's title to its images.

Grab them then, and you've got a chance...mess up, and they're gone.

ABSTRACT

One thing I do want to mention while talking about the concept of an "abstract". Your poster is the abstract for your research, just like all research papers have an abstract.

Your poster should NOT, on the other hand, *have* a section titled "abstract". Your poster IS the abstract, it IS the summary of your work. You don't need an abstract for your abstract. So keep that in mind.

The rule of thumb when it comes to scientific posters is… "Less is More!"

Believe me, I understand… You've been working on this research for a long time. You've put a hell of a lot of effort into it and it's important.

The impulse is going to be to cram AS MUCH information into that tiny poster as you can. RESIST THAT URGE!

Repeat after me…LESS…IS…MORE!

Your poster should have several sections, namely:

- Introduction

- Objectives

- Methods

- Results

- Conclusion

That's it! Each of those items should be its own little section, and each of those sections should have a short paragraph of information to go along with it. Sprinkle in a couple charts or graphs and maybe an image and you're done.

And remember, when it gets right down to it...your poster isn't really supposed to explain everything in great detail.

That's what YOU are there for. The poster is to let people know what your research is about and provide a few key points and the conclusion that you discovered.

Your poster is a tool that allows people to self-select themselves as being "interested" in whatever you're talking about.

Basically, it lets you know who is interested in your research (because they stopped walking and started reading the poster).

Now it's YOUR turn to jump in and start talking. YOU will fill in the details of your research, and the poster becomes secondary. You'll use it to point out key details while you talk about your research to whoever has stopped by.

You may even have a handout with more detailed information that you can hand out to the people who really seem interested.

THE ELEVATOR PITCH

There's a term in the world of business startups called the *elevator pitch*. When you've got an idea for a new company or business and you want to pitch it to potential investors, you develop an *elevator pitch*.

The basic idea is that you explain all your key ideas to a someone in the time it would take for you to ride in an elevator (it probably started with some overeager entrepreneur trapping a potential investor in an elevator…

You've got 60-90 seconds to pitch your idea before those doors open and the trapped investor can make a dash for freedom!).

Treat your poster the same way. Make it your elevator pitch. Better yet, you should use it as a tool IN your elevator pitch.

When someone wanders up to your poster at a conference and seems interested, give them your elevator pitch.

Spend 60-90 seconds summarizing your research. Engage in conversation with them, make eye contact, and use your poster as something you can point to from time to time to emphasize some point.

That's all.

In that scenario, you wouldn't want a poster filled to the brim with rows and rows of text. You want a few charts to point to, maybe an interesting image. And you want some basic statistics or data points that you can reference. That's it!

Less is more.

THE FLOW

The next thing to think about is the flow of your poster. Some people will just wander up and start reading it, and you might not have a chance to talk with them as they do (maybe you're talking to someone else already, or in the bathroom, or checking out the free drinks at the bar!).

Your poster needs to be designed to flow well, so that someone who reads the title can then easily discover where to start reading next, and how to follow the narrative of your poster in the way that you want them to.

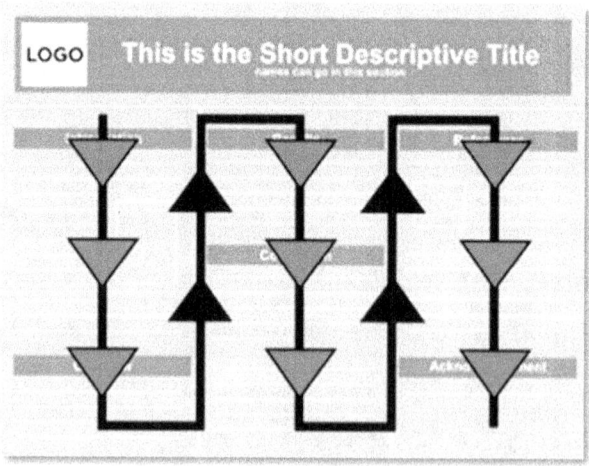

Some people get creative with their columns, expecting you to read top to bottom or left to right, and it can get confusing.

The easiest way to avoid this is to number each sub-head of each section. Start with 1..2..3..4..5..etc in the order that you want them read. Problem solved.

Also, many people tend to start the poster with their conclusions in sub-head number 1. Generally that's a matter of personal style, but personally I think it's a good idea.

Hit them with the conclusion right off the bat, then spend the rest of the poster explaining and supporting that conclusion. Why make people wait? Remember, posters are a drive-by type of thing.

You've just got a few seconds to grab someone's attention. Hit them with your big sell (your conclusion) right off the bat!

MARKETING

Like I mentioned earlier, your poster is ultimately just a big advertisement…a billboard…and the normal rules of marketing apply.

As someone who comes from a direct marketing background I can tell you that the headline is the most important component in any piece of advertising.

The headline (your poster's title) should do two things, and only two things.

1. Grab attention.

2. Get the reader to read the first sub-headline.

That's it! So let's talk about each of those things just a little bit.

The headline of your poster is the Title. Many people tend to make the mistake of treating the Title as an information transmission tool.

They think that since it's the one thing people will see as they wander by, they should try to transfer as much info through it as possible.

That's wrong.

Yes, your poster's title should convey the overall idea that your research discusses. But at the same time, its main purpose is to merely grab attention.

Remember - people are walking around, mingling, and probably enjoying the open bar. Your job is to get them to stop wandering around and look at your poster. The only way to do that is to grab their attention.

Don't be stupid about it. Don't grab attention just to grab attention. What I mean is; the title "FREE PORN!" will probably grab lots of attention but it's not the kind of attention you want.

Don't grab attention just to grab attention, keep it relevant to your research.

A great book that really dives into the science of headline creation is "Tested Advertising Methods" by John Caples.

You can grab a copy from Amazon for around twelve bucks and I recommend giving it a quick read. In it you'll learn things like how to use self-interest, news, or curiosity to draw people into a headline.

You'll learn the right and wrong way to create a headline. You'll learn some brainstorming headline writing techniques, and you'll learn thirty-five proven formulas for writing headlines.

Some of those techniques are more suited to selling products, but the general ideas behind the techniques can be applied to your poster.

I could write an entire book about headlines alone, and this isn't the place for a treatise on advertising so I'll leave it at that. But do think of your title from an advertising point of view.

The second purpose of your headline is to get the reader to continue reading your poster. Specifically, it's to get them to read the next sub-head, and then the body text underneath that sub-head.

That first block of body text should do one thing…convince the reader to read the NEXT sub-head and it's body text, which should convince the reader to read the NEXT sub-head and it's body text and so on and so forth until the reader has read your entire poster.

I mentioned FLOW earlier and that's what I was talking about.

Raise questions; then guide the reader to the answers. Make sure your poster is laid out in a non-confusing manner (number each sub-head so that the reader knows *exactly* how to proceed throughout your poster).

This seems like an obvious thing...but you'd be surprised how often people throw up roadblocks in a poster that interrupts a reader's progress and knocks the train off the tracks.

How do you know if your poster does a good job with its flow?

A good tip is to read your poster out loud yourself. Are there any sections where you stumbled as you were reading it?

Are there any words you had trouble pronouncing? Get rid of them! Keep editing until you can read the entire poster out loud without stumbling, without stuttering, without stopping for any reason.

Then ask a friend to read it out loud and watch them as they do. Did they get tongue-tied anywhere? Did they have to re-read any words? Did it flow smoothly?

Reading your poster out loud is a powerful tool that will show you exactly and immediately whether or not your poster flows well. This tip works whenever you write anything, be it a paper, report, book, article, or anything else.

And don't forget whitespace! Numbering your sub-heads is important, but using enough whitespace is equally important.

Remember, don't try to cram as much info into your poster as you can. Flow is much more important, because YOU are there to fill in the gaps if someone wants more information.

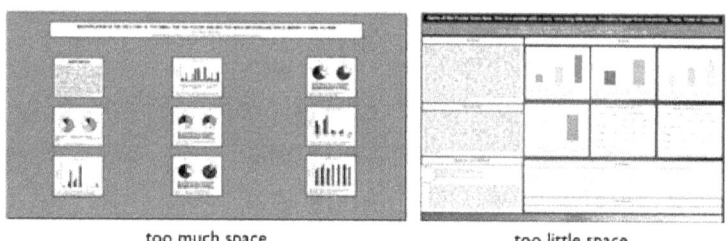

too much space   too little space

You should also have a handout that you can give to people that explains things in more detail. It should also point the reader to your website and contact information.

TARGET AUDIENCE

Throughout this whole poster-making process, make a point of keeping in mind your target-audience. Who are you trying to appeal to?

What sorts of things would appeal to those types of people? What would they expect to see? Have you provided it? Have you appealed to them?

Keep your target audience in mind when creating your headline. Keep them in mind when selecting charts, graphs, and images. Keep them in mind while writing your body copy and sub-heads.

In short…ALWAYS KEEP THEM IN MIND!

Just like a marketer crafts an advertisement to appeal to a specific demographic, you should craft your poster to appeal to the type of person who you're trying to connect with.

Every decision you make about your poster should be done with them in mind.

With every decision you make, ask yourself if it will make it easier or harder for your target audience to understand the key points you're trying to convey. Put yourself in their shoes; look at it from their point of view.

You are not important. What *you* want is not important. What *you* think is not important. It doesn't matter if YOU like the look of that background image…will your target audience like it?

Will it make it harder for them to read your poster or easier? Make every decision with them in mind.

And remember… they're moving… walking… wandering around – free drink in hand. How will you grab their attention and what will you do once you get it?

If you look at things from that point of view…you really can't go wrong!

CONCLUSION

Congratulations! You made it all the way through the book in one piece! I told you we could learn everything you needed about making a great research poster in less than a hundred pages! It wasn't that bad at all!

The key things I hope you'll take away from this quick and dirty guide are:

1. Use PowerPoint to make your poster

2. Use one of the free templates online (search Google) or at our MakeSigns.com site.

3. Stick to basic colors and resist using busy background images

4. Use charts and/or graphs and/or images to convey key info

5. Use dark text on a lighter background

6. Start with your conclusions in the first subhead, then support it with the remaining subheads

7. Whitespace is your friend, use lots of it!

8. Number your subheads in the order you want people to read them

9. Treat your Title like an advertising headline to grab attention and draw a reader in

10. Prepare your elevator speech before hand, use your poster as a supporting tool for the speech, not the other way around

11. Read your poster out loud to test for flow

12. Have your poster printed by a professional sign company to make sure it looks the best that it can look. Peace of mind is important!

I hope you enjoyed this book, and if you have any questions about anything you read here or anything else relating to poster creation or printing, please drop us a line! You can find us online at **MakeSigns.com**

You can talk to one of our Graphicsland design reps with our online chat tool, or you can fill out the contact form on the site, or you can give us a call toll free 1-800-347-2744.

We've also got a new online forum where you can post questions as well. Check it out here: http://forums.makesigns.com

A Quick Favor!

Can I ask you for a quick favor? If you enjoyed this book, would you drop me an email at **johne@graphicsland.com** and let me know, or better yet…head over to Amazon.com and leave a review.

Nothing helps promote a book more on Amazon than customer reviews and I'd really appreciate if you'd leave one! If this book was helpful, please take a moment to return to **Amazon**.com and leave a review here:

http://www.amazon.com/How-Make-Scientific-Research-Poster-ebook/dp/B00HUZUHVA

Or if you'd rather just drop me an email, I'd love to post your comments on our own website.

Also, if you've got friends or colleagues who you think might be interested in this book, please pass it along to them!

We offer bulk copies at heavily discounted prices if you think others in your department or organization might find the book useful.

Just drop me an email at the address above and ask me about prices.

Thanks again, and I hope you enjoyed the book!

-John Elder

CHAPTER FOUR

Resources

Here's a list of the different links that I mentioned throughout the book…

MakeSigns.com – Our poster printing website. Our company is called Graphicsland, Inc. and we print scientific research posters, store signs, general signs, and even bumper stickers.

Scientific Poster Forum – Our online forum where you can post questions and get answers about all things related to making a research poster.
(http://forums.makesigns.com)

Free PowerPoint Templates – Our free PowerPoint templates that you can download and use at absolutely no charge.
(http://www.makesigns.com/SciPosters_Templates.aspx)

PowerPoint Size Converter Tool – If you need to resize your original PowerPoint dimensions and need to change the aspect ratios, use our free tool.
(http://www.makesigns.com/SciPosters_PageSizeConverter.aspx)

Poster Size Calculator – This free tool will show you the different sizes your original poster can be converted to using the same aspect ratios.
(http://www.makesigns.com/SciPosters_PageSizeCalculator.aspx)

Sci-Poster Tutorial – This is our free online tutorial that goes into a little more detail about some of the topics we talked about throughout the book.
(http://www.makesigns.com/tutorials/poster-sizing.aspx)

Color Combo PDF – Download our free color combo pdf to help you pick your color palette.
(http://forums.makesigns.com/index.php?app=core&module=attach&section=attach&attach_id=2)

https://kuler.adobe.com/create/color-wheel/ - A free online color wheel tool.

http://www.colourlovers.com/ - Another free online color wheel tool.

http://colorschemedesigner.com/ Yet another free online color wheel tool, I like this one the best!

http://flikr.com/groups/pimpmyposter - Check out other people's posters and post your own to get feedback from the people of this neat Flikr group.

http://f1000.com/posters/browse?docTypeSearch=Poster – See tons of past posters that other people have created. Get those creative juices flowing!

Sci-Poster Price Comparison Chart – Check out the prices of our competitors with this comparison chart. (http://www.makesigns.com/SciPosters_PriceComparison.aspx)

THE END

Published By MakeSigns.com
Chicago, IL

Copyright © Graphicsland, Inc.
http://www.MakeSigns.com

All rights reserved. No part of this book may be reproduced or transmitted in any form or by any means without permission in writing by the editor or publisher, except when used by a reviewer in advertisements for this book, or other books or products by the author.

"This book is sold with the understanding that the publisher and author are not engaged in rendering legal, accounting, or other professional services, and is not intended to take the place of such services or advice. If legal advice or other expert assistance is required, the services of a competent professional person should be sought"

--From a declaration of principles jointly adopted by a committee of the American Bar Association and committee of the Publisher's Association